04/02/15

15/04/15

06 APR 2016

D0130628

Rob Waring, *Series Editor*

HEINLE
CENGAGE Learning™

Australia • ⬤m • United States

91120000164475

Words to Know

This story is set in the United States. It takes place in the central plains states of Kansas, Texas, and South Dakota, as well as other regions.

 Strange Weather. Read the paragraph. Then write each <u>underlined</u> word next to its definition.

Tornadoes are powerful storms that can have a wind velocity of up to 300 miles per hour and huge paths of destruction. Unfortunately, these storms remain some of the least understood phenomena in nature. Meteorologists often study tornadoes in order to try to predict when and where they will happen. If scientists can accurately forecast these dangerous storms, they can develop early warning systems and avoid loss of life.

1. people who study weather for their jobs: _____
2. fierce, fast-moving, circular wind storms: _____
3. unusual occurrences or happenings: _____
4. know about an event before it happens; predict: _____
5. speed, such as that measured in miles per hour or feet per second: _____

path of destruction

B **Tornado Chase!** Read the paragraph. Then match each word or phrase with the correct definition.

Tornado expert Tim Samaras [səmɛərɪs] and his team have a special mission. They want to put a high-tech instrument called a 'probe' into the center of a tornado. By deploying the probe into the storm, the team can measure the barometric pressure, wind speed and direction, temperature, and humidity inside it. This data will give them valuable information about tornadoes and how they are created.

1. mission _____

2. high-tech _____

3. probe _____

4. deploy _____

5. barometric pressure _____

6. humidity _____

7. data _____

a. the force of the air around earth

b. move equipment into a position for use

c. an important task or job

d. the amount of wetness in the air

e. information collected for analysis

f. using the most advanced technology

g. a scientific instrument used to explore the inside of something

tornado

1 mile = 1.61 kilometers

1 mile

Some of the more dangerous tornadoes can have a path of destruction over a mile wide.

Every year, more than a thousand tornadoes occur in the United States, often with deadly results. With wind velocities of up to 300 miles an hour, and billion-dollar paths of destruction, tornadoes are some of the least understood phenomena in nature. Many people lose their homes to these powerful storms each year—and some even lose their lives.

For many years, meteorologists and scientists have been trying to understand how and when tornadoes form. They have been doing this in order to develop better warning systems that will make people aware of the danger quickly. These more accurate warning systems would give people sufficient time to escape the extreme **violence**[1] of the storms before the tornadoes hit. However, to create these warning systems, scientists need to know when and how tornadoes begin. That's why tornado expert Tim Samaras and his team chase these incredibly strong storms. They travel straight into the action when everyone else runs away!

Samaras and his team of 'storm chasers' cannot prevent tornadoes from happening; however, their goal is to be able to help scientists get a closer look at these unpredictable storms. Samaras hopes that getting scientific information from the very center of the storm itself will help scientists to develop new methods of forecasting tornadoes. That way they can better know when and where dangerous tornadoes will happen.

---

[1]**violence:** actions done with force or causing great damage

 CD 2, Track 09

To get the data they need, Samaras and his group travel around the central part of the U.S. looking for tornadoes. They do this with the help of a number of high-tech **tracking devices**.[2] These devices are designed to help the team to get closer to tornadoes than anyone has ever been before. In order to follow the storms, Samaras must constantly watch all of the equipment, which can make it hard to stay focused. However, he knows he has to keep his mind on his task because chasing tornadoes can be extremely dangerous. This is especially true when the team gets caught in a storm with **hailstones**[3] that are almost two inches wide!

Samaras and his team are courageous in the way they chase tornadoes. While the average storm chaser watches and takes pictures from two or more miles away, Samaras and his group head straight for the heart of the action. They want to get as close as they can to the part of the storm that produces the tornado, which is sometimes called 'the bear's cage.' Samaras explains that he must enter this dangerous area because of a specific, scientific mission: he wants to place special weather probes directly in the path of the storm. These probes will then measure what is happening at the core of the storm itself. He talks about his objective: "My mission is to get right up close to it, get in its path, [and] deploy various pressure probes." Samaras also indicates that being so close to so much action can be quite exciting—and stressful. "So, you know," he says, "my **adrenaline**[4] is flowing."

---

[2] **tracking device:** a piece of equipment designed to follow the movement of something
[3] **hailstone:** rain that has frozen into a ball of ice
[4] **adrenaline:** a natural chemical produced by the body in times of danger or anger that makes one ready to fight or run away

With or without adrenaline, implementing the team's plan is not an easy job. Getting a weather probe into the center of a tornado is a lot of work. First, Samaras must find the tornado. As soon as he learns that a tornado is on the move, he drives out and chases it until he gets close enough to deploy his equipment. In order to do this part quickly, he keeps six weather probes—each weighing 45 pounds—ready to go in the back of his car. These probes are packed with instruments that measure barometric pressure, wind speed and direction, temperature, and humidity. Any measurements acquired by these probes from the center of the tornado are of great value. They could help scientists better understand how tornadoes form and how they maintain their strength.

There are negative aspects to tornado chasing, though. Once Samaras gets close to a tornado, he hardly ever gets to see it. He admits that he's usually so busy with the probes that he doesn't actually have much of an opportunity to really observe the storm. "So a lot of the times, because of what I'm doing … " he says, "I don't get a chance to really look out and see that tornado, just **appreciate**[5] its beauty." For that reason, it's helpful for Samaras to have a photographer with him on a tornado chase. Luckily, in this case, it's National Geographic photographer Carsten Peter.

---

[5]**appreciate:** understand the value or importance of something

Predict

Answer the questions. Then check your answers on pages 11 and 12.

1. Why does Peter want to photograph a tornado chase?

2. What challenges might he have in taking photographs during one?

3. Why might riding in a car be a problem for him?

This is a rare and unusual opportunity for Peter to take photos of a tornado close up. But why did he want to become involved in tornado chasing? He explains that for him, as a photographer, it was something different and exciting. He also wanted to be part of a mission that was attempting to discover new things. "It was important [to me] to be in the field with the scientists and to be on the **frontiers**[6] of science," he says.

Apparently, Peter takes the idea of being on the frontier and trying new things very seriously. He now wants to try a completely different approach to photographing one of these powerful storms. He thinks it actually might be possible to take pictures from the inside of a tornado. However, this kind of a task requires quite a bit of planning and preparation. First, photo engineers at National Geographic design their own photographic probe and fill it with still cameras and video cameras. Then, the designers check and recheck the probe to be sure that it is constructed very carefully and that its components can survive being inside one of these violent storms. At last the photographic probe is ready, but deploying his probe is just one of the challenges Peter faces.

---

[6]**frontier:** an unknown area in a field of knowledge

National Geographic photographer Carston Peter holds a photographic probe.

Another problem Peter faces is getting the shots that he needs using his regular camera. As he races along in Samaras's car, he comments on the chase: "For me as a photographer, the chasing is quite a big problem because we are sitting all the time in the car. It's very speedy. We have to catch up with the clouds, and we have to **zigzag**[7] around the clouds. And it's sometimes difficult to shoot, or to get some time for shooting." He then explains that the average speed of the car is between 60 and 80 miles an hour when he shoots photos. This can certainly make it very difficult to get a good shot.

Later, Peter explains how he felt as he raced across the countryside looking for storms, "The hardest [part] for me was to drive in [a] situation when I pass[ed] the most unbelievable pictures. Lights and storm situations I [have never] seen before, [and] I would have liked to photograph." But of course it was impossible for the team to keep stopping to take photographs. Peter continues, "You [had] to pass them because there was the likelihood that a tornado could form and we [had] to be [at the target] at the right time." During a chase, taking beautiful photographs is one thing, but the most important element of the mission must always be the tornado.

---

[7]**zigzag:** move from side to side in a generally forward direction

As they drive around the countryside looking for a tornado, Samaras and the team think about what will happen once they find one. If they do manage to set down their probe directly into a tornado's path, will the device be able to record any internal information or will it be destroyed by the power of the storm? At this point, nobody knows, but for the team, these kinds of questions are secondary. They must first worry about other things, like: is a tornado going to happen at all? Everything depends on nature—and nature is often unpredictable.

The formation of tornadoes is all a matter of changes in the cycles of air pressure and flow. In the spring, warm humid air that is rushing up from the south sometimes **collides**[8] with cooler dry air pushing down from Canada. When this happens, fierce storms called 'super cells' can be formed. When these super cells collide, conditions are ready for tornado formation. However, there are a number of variables about these conditions that remain unknown. Scientists still aren't certain exactly what aspects of these conditions cause a tornado to form. And, although they generally form under the same weather conditions, tornadoes vary greatly; they come in all shapes and sizes.

---

[8]**collide:** come together violently

Tornadoes are not rare in the U.S. Approximately 1,000 tornadoes touch down, or reach the ground, every year. Of these, the majority of them form in a part of the central plains area known as 'Tornado Alley.' This region, which is made up of about six states, has experienced some of the most powerful tornadoes in history. Because of this, it seems that it shouldn't be too difficult for Samaras and his team to find one. However, the difficulty lies in being able to be in the right place at the right time. Samaras explains: "First, you [have] got to have a storm. Then you [have] got to have a good-to-excellent road network of being able to get there. If you can't get there, it's no use."

Carsten Peter also stresses the difficulty of positioning the team and equipment near the center of a tornado. "It's the impossible **assignment,**[9] he argues. "You are trying to do something which is, in a way, impossible." For Samaras, the most difficult aspect of the mission is estimating exactly where the storm is going to hit. "Certainly, it's difficult at best," he says. "Maybe like two days before you can kind of narrow it down to maybe a state, like Kansas. And then maybe the day before, [we'll know that] it looks like maybe north central Kansas. And so by mid-to-late afternoon, we'll have it down to a specific town or maybe within a twenty-five-mile **radius.**[10] We average maybe anywhere between five hundred to a thousand miles a day driving." To emphasize just how much driving this is, he points out, "That means I change my oil every four days."

---

[9]**assignment:** a task or duty given to someone to do
[10]**radius:** circular measure of distance

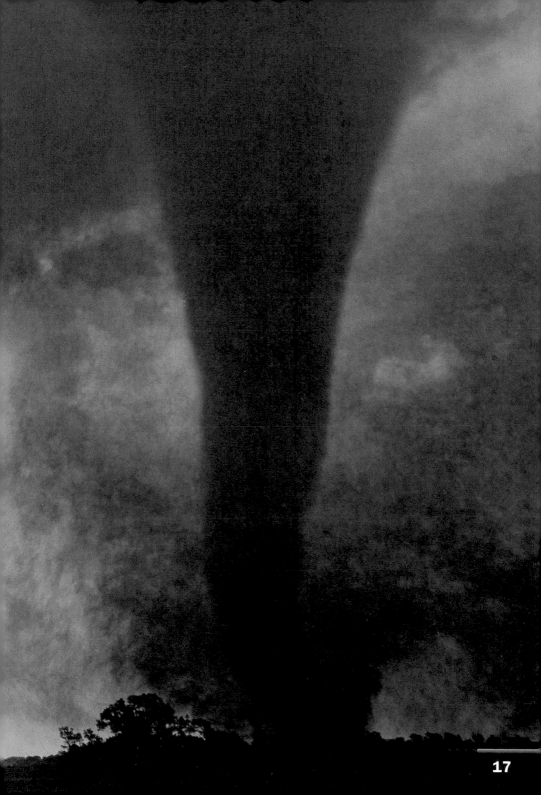

The team first set out in early May towards Texas, where a promising set of conditions for a tornado existed. Over the next six weeks, they drove over 25,000 miles as they traveled across eight states searching for the perfect location for a tornado. The weeks were filled with very long days and nights and there was a lot of time when the group just sat around, waiting for something to happen.

As the team waits in an open grass field for the next storm opportunity to come, Samaras talks about the waiting that is involved with tornado chasing. "We generally spend hours out in a field like this waiting for storms," explains Samaras, "just, you know, the wind blowing; it's hot and it's sticky and humid." He then continues, "[We're] watching clouds go up and there are days we're trying to watch storms, [and] sometimes they don't form. We'll be out there for hours. We'll watch the sun go down, no storms." This must be one of the most **frustrating**[11] parts of being a storm chaser.

---

[11]**frustrating:** annoying and difficult

Another important part of being a storm chaser is always being prepared. During tornado season, things can change very rapidly—and that's just what happens for Samaras and the group. As they relax, they suddenly get a report of a tornado nearby and everyone must move quickly! Carston Peter jumps out of the car, very happy that there are photographs to take after all that waiting. "[I] have to work. It's great!" he says excitedly.

For Samaras, this could be the most important moment of his job. He needs less than ten seconds to place the probe, turn the switch on, make sure the probe is facing north, and then run quickly back to the car. He's ready—but unfortunately, the storm isn't. At the last minute, the tornado moves away and Samaras and his team are frustrated once again. More often than not, this is the outcome when tornado chasers are waiting for a storm to hit out in the field.

Tim Samaras places a probe in the path of a tornado.

Tornado chasing isn't always so frustrating, however. Sometimes, everything comes together and the team is in the right place at the right time. At one point during the chase, Samaras takes the time to talk to reporter Lisa Ling. As he does so, he recalls a recent tornado in Manchester, South Dakota, which was an exciting experience. "We'd just arrived in the area around six p.m. local time. And we heard of the tornado warning, and off to our west, probably about fifteen miles, we could actually see what we call the 'developing wall cloud.' And we knew that this storm was about to produce a tornado."

In Manchester, the storm chasers were lucky. They managed to position their probes just minutes before a huge tornado hit. As he tells the story, Samaras shows the actual probe that survived the storm. It is slightly damaged, but still **intact**[12] and working, despite the fact that it was hit by a tornado that was a mile wide! "So this is the actual probe that was used?" asks the reporter. "Yes, this is it. This is probe three," replies Samaras as he points to the small, red, electronic instrument. "You can actually tell [that] this side of the probe was actually the one that was facing the tornado. And as you can see, there [are] lots of **debris pocks**[13] where who knows what actually impacted the probe." Ling is amazed that the probe survived and says, "Everything else in the path of that tornado was destroyed—including houses—but somehow this remained intact and functional." To this statement, Samaras replies with a simple, "That's right," as both people realize just how incredible that fact is.

---

[12]**intact:** whole; in one piece
[13]**debris pock:** [dəbri pɒk] a mark or slight hole made by the remains of something broken

Fact Check: True or false?

1. Samaras's team arrived at the Manchester tornado in time to deploy the probe.

2. The tornado was very wide.

3. The probe was completely destroyed.

4. The reporter is not impressed with the story.

The probe that survived the Manchester tornado provided Samaras with some very interesting new data. With that data, a surprising and important discovery was made—the barometric pressure inside a tornado drops farther than anyone had ever realized. This information may help to explain how some tornadoes are able to maintain their strength.

The group was able to deploy the scientific probe during the Manchester tornado and gain some valuable information, but what about the photographic probe? Was Carsten Peter able to take any photos of the inside of the storm? The storm did pass almost directly over Peter's camera probe. Unfortunately, though, it only recorded images from the edge of the tornado, but none from the inside. Still, Peter is determined to try again. It is now a kind of **fascination**[14] for him. He explains: "The closer I come to that type of natural phenomena, the more you **detect**,[15] the more you can discover, the bigger the fascination."

[14] **fascination:** something that powerfully holds one's attention
[15] **detect:** uncover or find

Where does this fascination come from? What is it about tornadoes that keeps some people coming back to chase them? In the same interview, Lisa Ling asked Samaras a similar question. "You've chased maybe a hundred or more tornadoes," she says. "Is it still as exciting?" For Samaras, it seems that it is the unpredictability of tornadoes that makes them so fascinating. "Absolutely," he replies, "because that's the beauty of it. Every day, you don't know what's going to happen. What kind of tornado are we going to see? What kind of conditions are going to exist? What's going to happen today? You may have had a frustrating taste the day before, but with the new day **dawns**[16] a new set of circumstances; another chance for some terrific storms!" For most people, tornadoes bring fear and destruction. However, for these tornado chasers, the amazing storms bring just what they need; an opportunity to learn, a chance to save lives—and a lot of excitement!

---

[16]**dawn:** come up; begin

Summarize

Imagine that you are either Tim Samaras or Carsten Peter. Write an e-mail or letter to a friend in which you describe your tornado chase.

After You Read

1. What is the main reason why Tim Samaras chases tornadoes?
   A. to show pictures of tornadoes to the public
   B. to document the damage done to homes
   C. to gather data for meteorologists
   D. to prevent tornadoes from happening

2. What is the most dangerous part of the tornado called?
   A. the hailstone
   B. the bear's cage
   C. the heart of action
   D. the path of storm

3. How does a probe usually help scientists?
   A. It locates tornadoes.
   B. It takes photographs.
   C. It chases storms.
   D. It collects information.

4. Which word in paragraph 2 on page 11 means 'not moving'?
   A. powerful
   B. still
   C. constructed
   D. violent

5. What's the hardest part about tornado chasing for Carsten Peter?
   A. He misses opportunities for amazing photographs.
   B. He doesn't like sitting in cars for a long period of time.
   C. The lights in the sky are difficult to photograph.
   D. The tornado might damage his equipment.

6. In paragraph 1 on page 15, 'it' in the phrase 'will it be destroyed' refers to a:
   A. tornado chaser
   B. probe
   C. tornado
   D. car

7. A super cell forms when warm humid air from the _____ collides with cooler dry air from the _____.
   A. east, south
   B. west, north
   C. east, west
   D. south, north

8. Why does the writer give details about Tornado Alley?
   A. to provide background information
   B. to offer an expert opinion
   C. to teach about tornado formation
   D. to stress how simple chasing tornadoes can be

9. Which of the following is NOT a suitable heading for page 20?
   A. Tornado Suddenly Forms
   B. Photographer Ready to Shoot
   C. Samaras Finally Turns On Switch
   D. Chasers Frustrated Again

10. On page 22, the word 'recalls' means:
   A. consults
   B. remembers
   C. displays
   D. reviews

11. The probe in Manchester provided information about _____.
   A. the prevention of tornadoes
   B. which direction a tornado will go
   C. the origin of a tornado
   D. how tornadoes last a long time

12. What does the writer probably think about tornado chasers?
   A. They attract danger.
   B. They love nature.
   C. They like the unknown.
   D. They have no fears.

CATCHING A *Tornado* —ON FILM!

F or years, Tim Samaras and his team have been chasing tornadoes to learn how to more accurately forecast them. They place high-tech weather probes in the paths of tornadoes and then go back later to collect the information the instruments recorded. However, on June 11, 2004, Samaras and his team were able to complete an even more exciting and scientifically important mission: they successfully placed a photographic probe containing five cameras directly in the path of a huge tornado. This enabled them to illustrate exactly what happens inside of one!

These photographic images have proven to be an incredibly important tool for measuring wind velocity in the lower parts of a tornado. This is especially important because this is the part of the storm which actually does most of the damage on the ground. Scientists have long used weather forecasting instruments to estimate the velocity of winds moving throughout the upper part of a tornado. However, they have never been able to use these instruments to study the bottom nine meters of a tornado as it races along its path of destruction. The winds in this area of the storm are the strongest and shift direction the most frequently. Being able to study this section in detail allows engineers to design buildings that have a better chance of surviving a fierce tornado.

The images from Samaras and his team are particularly useful because they provide a series of pictures taken only 16 one-thousandths of a second apart

Samaras places the probe in the path of the tornado.

The tornado approaches the probe.

A leaf flies through the air at over 200 kilometers per hour.

The tornado rolls a piece of farm equipment past the camera.

These four images taken by the probe show various stages of a tornado.

(.016 seconds apart). Samaras and the other scientists used the measurements of Samaras's body in the picture frame to determine the size of objects in the photos. By tracking objects as they moved through the timed photos, the scientists were then able to accurately determine the velocity of the objects. They did this by measuring how far the objects moved in a specific amount of time.

During their first study, they found that a stick shown in one of the pictures was traveling at almost 115 kilometers per hour. They also discovered a leaf that was flying along at 200 kilometers per hour!

CD 2, Track 10

Word Count: 349
Time: _____

Vocabulary List

adrenaline (7, 8)

appreciate (8)

assignment (16)

barometric pressure (3, 8, 25)

collide (15)

data (3, 7, 25)

dawn (26)

debris pock (22)

deploy (3, 7, 8, 11, 23, 25)

detect (25)

fascination (25, 26)

forecast (2, 4)

frontier (11)

frustrating (19, 20, 22, 26)

hailstone (7)

high-tech (3, 7)

humidity (3, 8, 15, 19)

intact (22)

meteorologist (2, 4)

mile (3, 4, 12, 16, 19, 22)

mission (3, 7, 11, 12, 16)

path of destruction (2, 4)

phenomena (2, 4, 25)

probe (3, 7, 8, 11, 15, 20, 21, 22, 23, 25)

radius (16)

tracking device (7)

tornado (2, 3, 4, 7, 8, 9, 11, 12, 15, 16, 19, 20, 21, 22, 23, 26, 26, 27)

velocity (2, 4)

violence (4, 11, 15)